The American Middle Class Is Being Reorganized:

Upper Middle Gradually Replaced by Degreed Immigrants,
Lower Middle by Millennial's

C. E. Nough

Table of Contents

Chapter 1: HOW THE AMERICAN ECONOMY WORKS

The top 2% of Americans, wealth-wise, could be characterized as money magicians. Their main, if not only, talent is shuffling "funny-money".

"Funny-money"? Since America went off the gold standard, the American dollar is "floating". The Federal Reserve prints money from nowhere whenever it feels the need, and the number of dollars in circulation in the American economy fluctuates.

American money is based solely on the Good Faith and Trust in the US Government. At this time, the World remembers the Great Recession which vaporized roughly one-third of their wealth. And the American people's confidence in Congress, at least, is at an all-time low.

It has been speculated that first people in America, and then Internationally, may lose confidence in the intrinsic value of the dollar.[1] This event, should it occur, would be far worse than the Greek disaster.

Nevertheless, these 2% magicians are diligently managing the American dollar and the American economy. They are clearly in control, and are currently the most vulnerable.

But, they are <u>dependent upon innovators to create new products</u>.

There are *Nouveau riche* inhabitants in that population NOW, such as Bill Gates and Steve Jobs. But they *were born into the Middle and Lower Classes.*

---

[1] See comments by Ron Paul, Ron Paul vs. Paul Krugman: *Austrian vs. Keynesian economics in the financial crisis.* or view the concise interview online:
https://orders.cloudsna.com/chain?cid=MKT147626&eid=MKT192494&snaid=&step=start##AST26387

[Bill] Gates was born in Seattle, Washington on October 28, 1955. He is the son of William H. Gates, Sr. and Mary Maxwell Gates. Gates' ancestral origin includes English, German, and Irish, Scots-Irish. His father was a prominent lawyer, and his mother served on the board of directors for First Interstate BancSystem and the United Way. – Wikipedia

And

[Steve] Jobs' adoptive father, Paul Reinhold Jobs (1922–1993), grew up in a Calvinist household, the son of an 'alcoholic and sometimes abusive' father. The family lived on a farm in Germantown, Wisconsin…. Clara [Job's adoptive mother], the daughter of Armenian immigrants, grew up in San Francisco and had been married before, but her husband had been killed in the war. After a series of moves, Paul and Clara settled in San Francisco's Sunset District in 1952. As a hobby, Paul Jobs rebuilt cars, but as a career he was a 'repo man', which suited his 'aggressive, tough personality.' Meanwhile, their attempts to start a family were halted after Clara had an ectopic pregnancy, leading them to explore adoption in 1955. – Wikipedia

How does the American 98% contribute? Through breeding out-of-the-box thinkers, and innovators. The wealthy only profit from selling "new" and "better" stuff. Here in America, and increasingly Internationally.

The process necessary to develop these innovators is *self-actualization, the achievement of one's full potential.* Such Americans in the past have produced a constant stream of innovations that are profitable.

How does one become self-actualized? By climbing a pyramid.

Abraham Maslow's famous Pyramid of Needs[2], leading to self-actualization, is shown below. *Caveat*: Maslow asserted that full

---

[2] https://en.wikipedia.org/wiki/Maslow's_hierarchy_of_needs

potential could only be reached by <u>attaining all of the lower levels of existence, FIRST</u>!

Basic human Needs

This is the historical pathway to innovation and *functioning* genius. So, what is limiting that 98% contribution today?

Using the financial ploy of exporting jobs (outsourcing and offshoring) and factories, Americans have lost stability in meeting their Basic Physiological Needs, <u>Maslow's level 1</u>.

Breaking unions, inability to pay promised retirement benefits, massive private and public bankruptcies, increased cost of access to health care, insufficient police coverage due to layoffs, etc. has destroyed confidence in security and safety, <u>Maslow's level 2</u>.

The whole supporting social structure that has produced America's innovators from the Middle Class of the past has been undermined!

What else?

We don't need to just maintain our population of innovators, but also need (have jobs for) to produce *more* innovators in the future to keep the American economy competitive.

The US Bureau of Labor Statistics has published projections of the work force growth expected through 2018 and up to 2024.[3]

The Table below demonstrates an overall 19% growth in **Science, Technology, Engineering, and Mathematics** (STEM) professionals 2008 – 2018 (in thousands);

Occupation	2008 jobs	2018 jobs	% change
All Occupations	150,932	166,206	+10%
Professionals	31,053	36,280	+17%
ALL STEM Occupations	5,667	6,747	+19%
Biology	279	354	+27%
Computer Science & Math	3,540	4,326	+22%
Physical Scientists	276	317	+15%
Engineers	1,572	1,750	+11%

The 2014 – 2024 projection in percent of STEM professionals is:[4]

Occupation	% change
Bio-Chemist/Physicist	+8.2%
Chemist	+2.6%
Computer and IT research	+10.7%
Geoscientist	+10.5%
Mathematics*	+21%
Physical Science	+7.9%
Civil Engineer	+8%
Mechanical Engineer	+5%
Environmental Engineer	+12.2%

*Note: the percent growth is large, but the actual number of Mathematician positions is small.

The Table above indeed shows projected growth through 2024, but the growth in STEM NEED has slowed down from a decade ago.

---

[3] *US Bureau of Labor Statistics Projections*. Retrieved from: http://www.bls.gov/projections/; http://www.itif.org/files/2010-refueling-innovation-economy.pdf

[4] Retrieved from: http://data.bls.gov/projections/occupationProj

One might speculate that the slowdown in need is related to offshoring and exporting STEM jobs, as opposed to an American decline in participation in STEM innovations.

Nevertheless, the American **growth in STEM professional**s has remained flat for two decades. Non-responsive.

Numbers below show *in thousands* the Science, Technology, Engineering and Mathematicians ... graduates:[5]

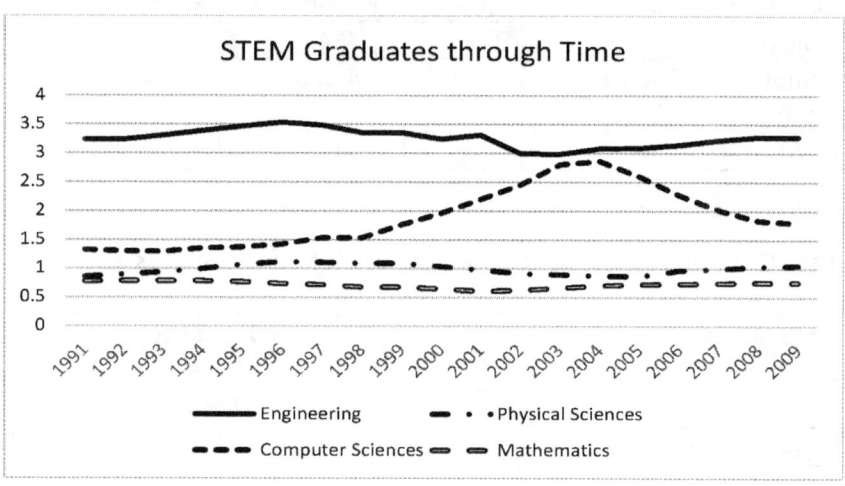

And World-wide, the rate of graduation in STEM areas shows America behind (in percent) of STEM degrees awarded globally:[6]

---

[5] *Science and Engineering Indicators 2012*. Retrieved from: http://www.nsf.gov/statistics/digest12/stem.cfm#3; FEBRUARY 2012, *Executive Office of the President, President's Council of Advisors on Science and Technology, Report To The President, Engage To Excel: Producing One Million Additional College Graduates With Degrees In Science, Technology, Engineering, And Mathematics*. Retrieved from:https://www.whitehouse.gov/sites/default/files/microsites/ostp/pcast-engage-to-excel-final_2-25-12.pdf; *As STEM Graduate Rates Decline, Funding and Improving STEM Education Is More Crucial Than Ever*. Huffington Post. Retrieved from: http://www.huffingtonpost.com/vivian-r-pickard/as-stem-graduate-rates-de_b_3744718.htmlhttp://flbog.edu/pressroom/newsclips_detail.php?id=31217
[6] *The global race for STEM skills*. Retrieved from: http://www.obhe.ac.uk/newsletters/borderless_report_january_2013/global_race_for_stem_skills ; Accenture Institute for High Performance.

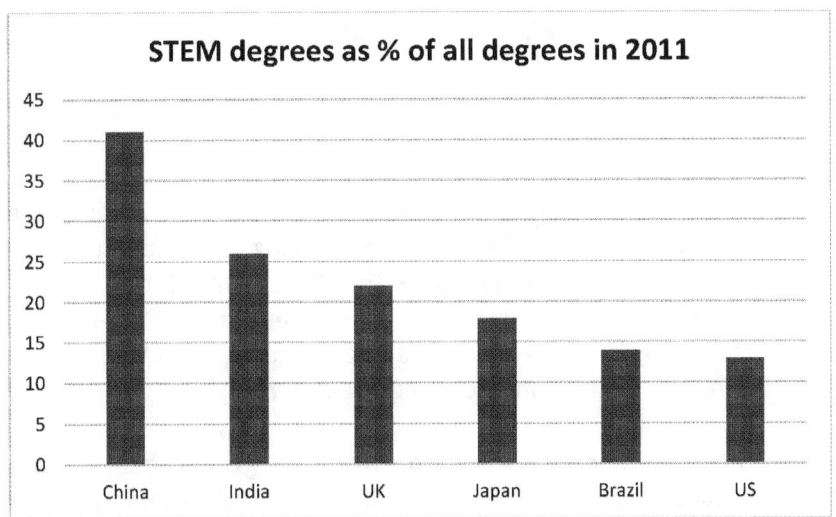

STEM degrees as % of all degrees in 2011

Note: The large Chinese and Indian populations contribute to % degrees disproportionately. Yet, these are numbers of graduates going into a creative profession!

Finally, what else contributes to America's slide in producing STEM graduates?

The Millennial (current) generation has a different psychological profile from past generations. <u>Millennial's, in general, are mismatched psychologically and philosophically to the Engineering and Science professions.</u>

This different philosophy isn't laziness, as often claimed, but is *a core-value conflict with the STEM profession attributes*, as well as others (expanded in Chapter 2). This personality mis-match is not new. When small farmers and the unemployed moved into cities and took Blue Collar jobs, there were mis-match problems the first generation. And, there have always been issues with someone moving from Blue Collar to White Collar employment.[7]

---

[7] *Limbo: Blue Collar Roots, White Collar Dreams*, Alfred Lubrano.

Evidence of the Millennial's response to STEM professions:

Loss #1: Studying the student pipeline into Engineering and Science fields reveals another problem for America.

> High school students aren't sticking with STEM. Even though the number of jobs in science and engineering is expected to surge in the years to come, close to 60 percent of the nation's students who begin high school interested in science, technology, engineering, and math, or STEM, change their minds by graduation, according to a report.[8]

> Overall student interest has been gradually climbing for about a decade, with about 1 in 4 of all high schoolers excited about pursuing a STEM major or career. But keeping many of those students attracted to such subjects is proving a challenge. 'Tying education to the workforce needs is critical to the future of the nation' ....[9]

There is a known reason for *part* of this change of mind: bad first science course experience.[10] Contributing, there is a shortage of qualified, trained teachers in first level courses. Second string teachers fill in the gap.

And this loss has probably more recently been exacerbated by poor education in Mathematics, upon which Science is based. The discouragement over these courses is a game-changer for Millennial's.

A related factor is that Millennial's are at this point in High School are investigating and choosing careers. Their psychological profile (expanded in Chapter 2) contains a core, enviable element of *balanced home-and-work time.*

---

[8] *Many High Schoolers Giving Up on STEM*, USNews. Retrieved from: http://www.usnews.com/news/blogs/stem-education/2013/01/31/report-many-high-schoolers-giving-up-on-stem

[9] from STEMconnector and college planning service My College Options

[10] *They are not dumb, they are different"*, Sheila Tobias.

When Millennial's find that the work week for Science and Engineering involves on average a 50 hour work-week[11] (for 40 hours pay) they move on.

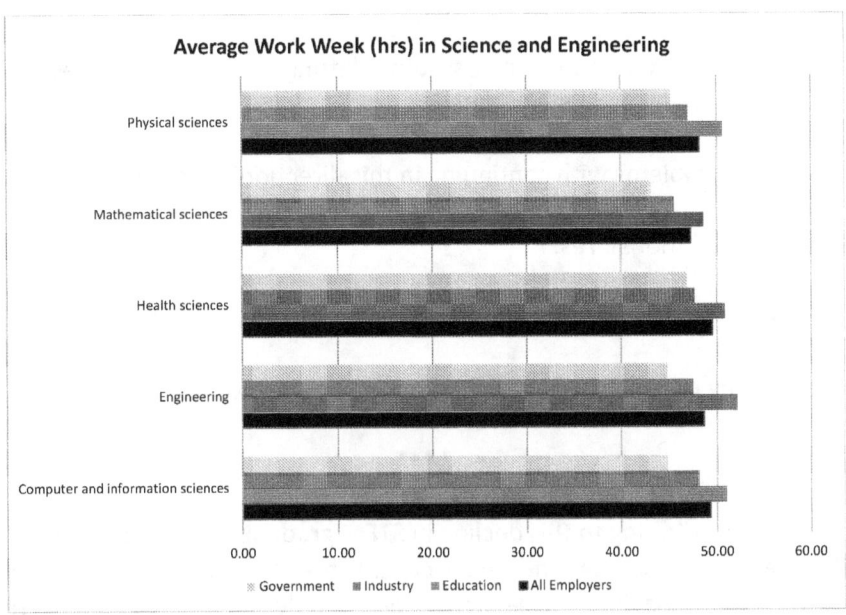

Average Work Week (hrs) in Science and Engineering

Categories (top to bottom): Physical sciences, Mathematical sciences, Health sciences, Engineering, Computer and information sciences

X-axis: 0.00, 10.00, 20.00, 30.00, 40.00, 50.00, 60.00

Legend: Government, Industry, Education, All Employers

Loss #2: There is another 60% loss[12] of STEM students in College/University. Similarly, this loss is both due to poor and uninspired Instructors. There are also faculty without knowledge of the appropriate and optimal pedagogy.[13]

No matter what STEM area, the first exposure to the subject should be taught by the BEST instructors in the Department. This strategy is only implemented in some Institutions and Departments.

There have been attempts to stimulate interest in the Science and

---

[11] *Top-Level Professionals View 40-Hour Work Week as Part-Time: Report,* Huffington Post. Retrieved from: Htpp://www.huffingtonpost.com/2011/06/30/americans-now-view-40-hou_n_888231.html

[12] *Encouraging STEM Students Is in the National Interest.* The Chronicle of Higher Education. Retrieved from: http://chronicle.com/article/Encouraging-STEM-Students-Is/132425/

[13] For Physics only, refer to http://www.colorado.edu/per/

Engineering fields, but I claim you cannot brainwash a person's psychological core easily, nor ethically.

Conclusion:

Is the <u>American</u> education system letting down the American technological needs pipeline? Yes!

Are there problems with continuing in this direction? Definitely.

Are there solutions? Yes.

Is America going through a paradigm change? Probably.

Is there any danger? Definitely. Overseas DEPENDENCE!

****

Groups contributing to this decline in STEM graduates are identified, not for blame, but so that they may be enlightened, and reconsider and adjust their behaviors. These are not bad people; they are just unaware of the big picture and the consequences of their actions from a larger viewpoint.

CHAPTER 2: WHERE IS AMERICA NOW?

Who is REALLY paying to operate the US Government?

The total federal <u>taxes collected</u> pie is shown below for 2013:[14]

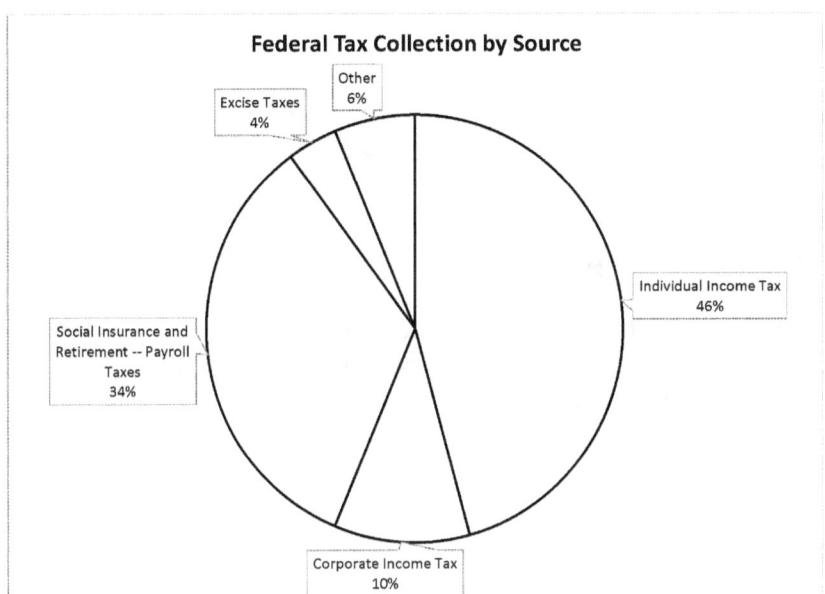

POINTS TO PONDER:

- Payroll taxes are not really Federal revenue. They pass through to entitlement programs such as Social Security and Medicare.

- Not counting these funds as spendable "revenue", the total tax contribution to Federal operating income by source becomes:

---

[14] https://www.whitehouse.gov/omb/budget/historicals

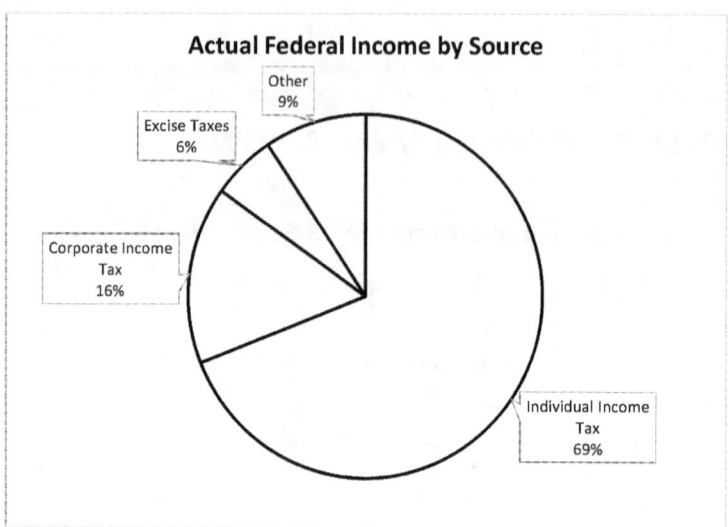

Actual Federal Income by Source

- Other 9%
- Excise Taxes 6%
- Corporate Income Tax 16%
- Individual Income Tax 69%

Personal Income Taxes

Using the IRS data set[15], one can plot actual income tax in dollars paid by income bracket. If one looks at just the ordinary taxpayers, it looks as if the rich are taking advantage:

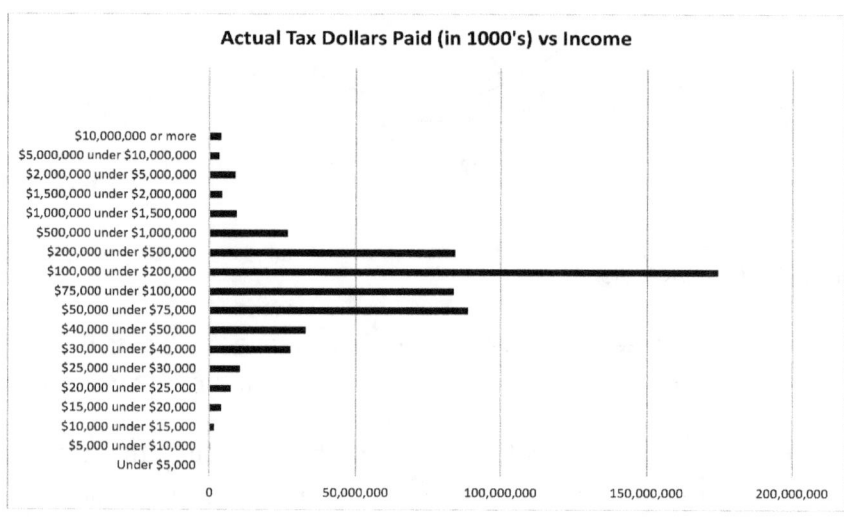

Actual Tax Dollars Paid (in 1000's) vs Income

[15] Internal Revenue Service. Retrieved from: https://www.irs.gov/uac/SOI-Tax-Stats---Individual-Statistical-Tables-by-Size-of-Adjusted-Gross-Income

However, if one includes taxpayers with Capital Gains tax, the distribution of payments looks like:

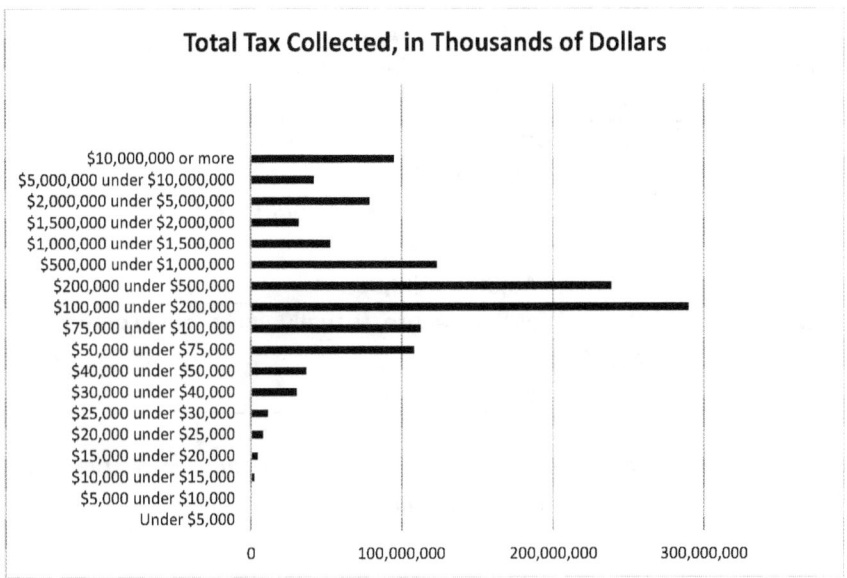

Total Tax Collected, in Thousands of Dollars

The conclusion is that the top 2% (income $500,000 and up) pay about ½ the individual income taxes and most of the corporate income taxes.

The rest of America is riding along.

Why are the wealthy carrying this burden? Clearly they have the money and political clout to exempt themselves!

The wealthy *need innovators* from the rest of America, and are willing to fund the Federal Government to manage the 98%, and to protect the "genius" nursery.

Reiterating that the origin of innovation and genius derives from the Middle Class, and granted freely that it is difficult to define the so-called economic classes, let's investigate the Middle Class.

Consider a traditional academic definition[16]:

[16] Thompson, W. & Hickey, J. (2005). *Society in Focus*. Boston, MA: Pearson, Allyn & Bacon;

Class	Typical Characteristics
Upper (2%)	Top-level executives, Celebrities, Heirs Income $500,000 and up
Upper Middle (15%)	Highly educated professionals and managers Income $75,000 - $500,000
Lower Middle (32%)	Semi-professionals and craftsmen Income $35,000 - $75,000
Working (32%)	Clerical pink- and blue-collar workers High school educated Income $16,000 - $30,000
Lower (14-20%)	Poorly paid, insecure, usually have government assistance Income $30,000 and below

Notice again that the source of innovation and invention has traditionally come from the Middle Class, which is shrinking.

Is Congress at all aware of what is going on with their constituents? Or are they caught up in their personal drama?

The American Government has become a place of "of the lobbyists, by the lobbyists, and for the lobbyist's clients".

How did we get here? Why has American Citizen been replaced by the Lobbyist?

Election finance abuse.

Consider a snake eating its own tail:

PROGRESSION:

1. The tail: A person wants to run for US Representative. Optimistically, because they think they can make a difference in their community, or maybe they want to use this position as a step up to a higher office. *Either path needs longevity in office to be fruitful.*

2. US Representative election campaigns are expensive. (Note that representatives only have to advertise in a District; Senators have to advertise in a whole State and their election costs are vastly more expensive.)

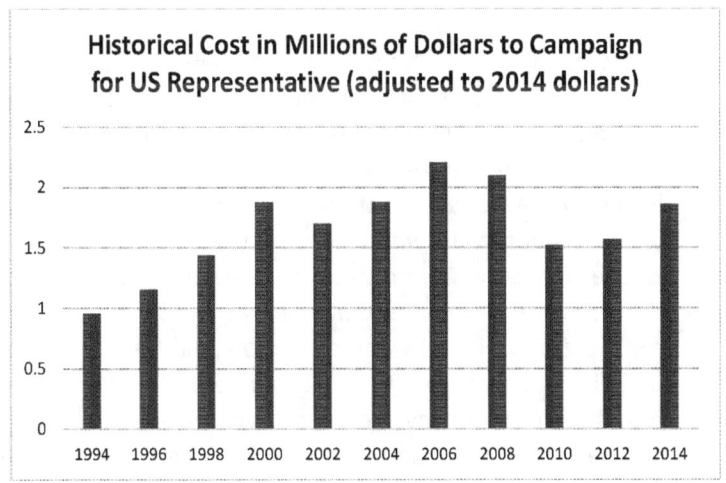

Historical Cost in Millions of Dollars to Campaign for US Representative (adjusted to 2014 dollars)

3. There is a choice of funding sources. 1) A work-intensive grass roots solicitation of donations (Barach Obama and Bernie Sanders have done well using social media to move forward with grass roots sponsored campaigns), or 2) several large donors.

   The problem with the second option (and the one chosen most often), is that large donors then have expectations, and will send lobbyists to remind Congresspersons of their debt.

   Bonus: these days, the attorney for the lobbyist will write most of your House bills for you.

   Dilemma: follow your plan to help your community (and lose your donor), or pay your debt.

4. In two years, you will have to campaign again. Little time to make a difference, as well as to seek funding and campaign again.
5. The snake-head eats the tail.

How about trying to break the cycle? One idea would be to have fewer elections that can be "bought". From someone who attempted to lessen corruption:

   This was my introduction, as a freshman member of Congress, to how power works on Capitol Hill: Following the lead of the President, a small group of us introduced a measure to extend the term of House members from two years to four. Given its support in the White House, we thought we had a chance for success, so we were optimistic when we approached the chairman of the House Judiciary Committee, an awesome and fearsome New Yorker named Emanuel Celler. I was designated the spokesman for the group. How, I wanted to know, did Mr. Celler stand on the bill? 'I don't stand on it,' he responded. 'I'm sitting on it. It rests four-square under my fanny and will never see the light of day.' He was right. It didn't. That day I learned a vivid lesson about congressional power: that some individuals [emphasis mine] have enormous power within the institution either to move legislation forward or to kill it.[17]

So much for the concept of equal representative government and equal representation of Congresspersons.

Considering the Constitutionality of these power-people:

> *The House of Representatives shall chuse their Speaker and other Officers; and shall have the sole Power of Impeachment.*
> **— U.S. Constitution, Article I, section 2, clause 5**

> The Speaker is the political and parliamentary leader of the House. The Constitution mandates the office, but the House and Speakers have defined its contours over time. Some Speakers *have aggressively pursued a policy agenda for the House while others have, in the words of Speaker Schuyler Colfax of Indiana, 'come to this chair to administer [the] rules, but not as a partisan.'* Regardless, the Speaker—who has always been (but is not required to be) a House Member and has the same duties to his or her local constituents like the other 434 Members—is at the levers of power. The Speaker is simultaneously the House's presiding officer, party leader, and the institution's administrative head, among other duties. [18]

Has anything happened to the Middle Class, other than their instability? Yes, the Millennial personality profile.

Yes. The CSI-type personality profile of the Millennial worker. (This Millennial generation is just entering the workforce.)

---

[17] *How Congress Works and Why You Should Care*. Lee H. Hamilton, Indiana Press (2004).

[18] *History, Art, & Archives*. United States House of Representatives. Retrieved from: http://history.house.gov/Institution/Origins-Development/Speaker-of-the-House/

Let's take a side trip into what personality is.

> **Personality** has to do with individual differences among people in behaviour [sic] patterns, cognition and emotion....

> The term "personality trait" refers to enduring personal characteristics that are revealed in a particular pattern of behaviour [sic] in a variety of situations. -- Wikipedia

Current research in Psychology indicates that 60% of your personality traits are locked in at conception, innate. The other 40% is developed as you have lived in your individual life-time environments.

> In psychology, **temperament** refers to those aspects of an individual's personality, such as introversion or extroversion, which are often regarded as <u>innate rather than learned</u>. A great many classificatory schemes for temperament have been developed; none, though, has achieved general consensus in academia. -- Wikipedia

> The **biological basis of personality** is the theory that personality is influenced by the biology of the brain. Though closely related to personality psychology, the biological basis of personality focuses on *why* or *how* personality traits manifest through biology, in addition to identifying personality traits. This is investigated by correlating personality traits with scientific data from experimental methods such as brain imaging and molecular genetics. -- Wikipedia

For reasons unknown (though probably due to connections in your DNA), the profession that will make you *most* satisfied for the *longest period of time* is directly related to your temperament, which is locked in at conception. As mentioned above, the two commonly used test for temperament type (and thus preferred profession) are the Keirsey-Bates and the Meyers-Briggs temperament sorters.

The Millennial Temperament:

Some things are said more succinctly about the Millennial generation than I can, so I quote some good researchers and writers:

Millennial's can be characterized by ...

<u>Self-belief caused by reinforcement and continual praise</u>

These Millennial workers were given certain messages by their parents and the other influencers in their lives, many of which influencers came from their screen based activities. This has created a self- belief system in which they see themselves as they desire to be. Sometimes those dreams are shattered in the cold hard light of reality. Staring continually at fictional actors and video game stars tends to create indelible impressions of misguided self -belief.

However, they [also] have been given examples, if only screen based examples, of being leaders and risk takers brimming with self- confidence and seeing themselves as goal orientated driven by success achieved in higher and higher levels in a video game.

The encouragement by their parents, helpers and being the most materially endowed and entertained generation has created a confidence to achieve, even at video game level, and bred a self -assured generation not afraid to speak their mind.

These Millennial's were indulged and encouraged to do numerous things and were possibly over praised for effort and for trying rather than praised for final accomplishment. They then learnt [sic] that praise was needed for every step along the way, not only for crossing the finishing line.

This has made them goal orientated ... [but with] the constant need for praise ... required to meet their goals....

Trust issues

They are trusting of others in their generation. However, they are but not so trusting of their parents' generation as they have experienced many broken promises and some have seen their parents laid off work after giving years of loyalty to their jobs at the expense of time with their children.

They have also inherited *inter alia*: huge government debt, major economic woes created by their avaricious older generation, and a youth unemployment rate of between 17% – 55% around the world. Coupled with that many are saddled with sizeable student loans which even bankruptcy can't void.

In the US alone, student loan debt is over $1 trillion and is higher than credit card debt from the entire economy. The doubting and mistrust of this generation towards the older generation, many of whom are employers and /or co- workers, is a major issue in learning how to manage and motivate this Millennial generation of employers, employees and entrepreneurs.[19]

And from another viewpoint:

Millennial's can be characterized by ...

Sheltered

Highly protected as children. Grew up in a time of increasing safety measures (car seats, baby on board signs, school lockdowns). They were rarely left unsupervised. They were sheltered from having to take care of their own conflicts as parents advocated on their behalf, and "spared" them from unpleasant experiences. As college students, they may expect faculty and staff to shelter, protect, and nurture them – and resolve their conflicts for them....

---

[19] *Get the Job*. Retrieved from: http://www.getthejob.co.za/psychological-profile-Millennial's/

Confident

They are motivated, goal-oriented, and confident in themselves and the future. They expect college to help launch them to greatness. They may brag about their generation's power and potential. They have high levels of optimism and they feel connected to their parents. They are assertive and believe they are 'right'. In Canada the Millennial generation is called the 'Sunshine' generation.

Team-Oriented

They are group oriented rather than being individualists. They may sacrifice their own identity to be part of the team. They prefer egalitarian leadership, not hierarchies. They are forming a tight-knit generation. While they are group-oriented within their own cohort, they may "politely" exclude other generations. They do not want to stand out among their peers, they want to be seen as part of the group. They dislike selfishness and are oriented toward service learning and volunteerism.

Achieving

Grade points are rising with this generation and crime is falling. The focus on getting good grades, hard work, and involvement in extracurricular activities, etc. is resulting in higher achievement levels. They see college as the key to a high paying job and success, and may miss the bigger picture of what a college education is all about. They are pressured to decide early on a career – and have been put on a career track orientation since grade school. Their focus is more on the world of achievement rather than personal development....

Pressured

[They are] tightly scheduled as children and used to having every hour of their day filled with structured activity. This generation may have lost a sense of pure spontaneous play. They may struggle with handling free time and time management in general. In elementary, middle, and high school, have had more hours of homework and less free time than any of the previous generations. They feel pressured to succeed. They've been pushed hard to achieve, to avoid risks, and to take advantage of opportunities. They may take on too much, and then think others should be flexible with them when they want to negotiate scheduling conflicts. They think multi-tasking saves time and is a smart thing to do, but aren't usually aware of the poorer quality of results.

Conventional

Respectful to the point of not questioning authority. They are civic-minded and believe the government knows what's best and will take care of them. They fear being considered non-conformist. Their clothing, music, and cultural markings will be very mainstream. They value their parents' opinions very highly. They support and believe in social rules, and are more in line with their parents' values than most other generations have been. They are trying to invite rules and norms back into the culture.[20]

Given this Personality Profile, what professions do Millennial's prefer?

I guess America expects this up-and-coming generation to be like themselves. Not so.

---

[20] Millennial's Go to College. (2003). Neil Howe and William Strauss

There has been a serious change in Millennials professional goals, to match the core values of Millennials. See Table below:

Job Title	Growth by 2022	Median Income	Millennial Share
Geological and petroleum technicians	15%	$52,700	40%
Dietitians and nutritionists*	21%	$55,240	36%
Medical scientists*	13%	$76,080	36%
Agricultural and food scientists	9%	$58,610	42%
Surveyors, cartographers, and photogrammetrists	12%	$56,530	39%
Fundraisers*	17%	$50,680	37%

Excerpted from a Table from *Time Magazine*[21]. *More than 50% female.

---

[21] *The 25 Most Promising Jobs for Millennials.* Retrieved from:
http://time.com/money/4013359/best-jobs-Millennial's/

Job Title	Growth by 2022	Median Income	Millennial Share
Physician assistants*	38%	$91,000	45%
Actuaries	26%	$94,000	57%
Statisticians	27%	$76,000	44%
Biomedical Engineer	27%	$87,000	43%
Computer Research Scientist	15%	$102,000	45%
Market Research Analysts	32%	$60,000	44%
Nuclear Engineer	9%	$104,000	60%
Elevator Install Repair	25%	$77.000	41%
Petroleum Engineer	26%	$77,000	35%
Therapists*	27%	$70,000	37%
Dental Hygienists*	33%	$70,000	37%
Logisticians	22%	$73,000	37%
Financial Analysts	16%	$72,000	41%
Software Development	19%	$87,000	36%
Pharmacists*	15%	$117,000	35%
Public relations and fundraising managers*	13%	$95,000	35%
Public relations specialists*	12%	$54,000	44%
Credit Analysists	12%	$54,000	44%
Agents and business managers of artists, performers, and athletes	10%	$63,370	42%

Note: a large number of these preferred positions are para-professionals. Less responsibility, more flexible time, more teamwork, more supervision (a source for praise).

The Millennial personality profile and *this* list of professions seem to correlate.

What jobs do Millennial's Hate?

Quoting:

> To help those business owners better understand what Millennial's want from their work, here are six types of jobs Millennial's don't particularly care for.
>
> ### A dead-end job
>
> A new study says that about two-thirds of today's young people are afraid that their current job will lead to a dead end, according to *The Standard*.
>
> The survey, which interviewed 2,000 people ages 15 to 22, found that youngsters are also worried that they won't find a job that fits their creative desires. In fact, youngsters are more likely to put off some jobs because it doesn't fit with their passion.
>
> There are some fantastic opportunities for both women and men in these sectors, so I'm concerned to hear that so many young women are put off by careers in science, technology, engineering and maths [sic], Claire Miles of British Gas told *The Standard*.
>
> ### The 9-to-5 job
>
> My colleague Mandy Morgan wrote in late July that Millennial's often avoid standard 9-to-5 jobs because they much

more prefer an occupation that allows them to serve a greater purpose through work.

In fact, 72 percent of current Millennial students prefer a job with a greater purpose, Forbes reported. And 71 percent of Millennial's even want their co-workers to be their second family, showing that they want their job to be more than just a place that pays them, Forbes reported. Rather, they want a meaningful experience from their work.

A job with too much work, not enough life

The Washington Post's Brigid Schulte reported in May that Millennial's want work-life balance from their jobs, especially as they struggle to balance their jobs and their families — something that older generations have had no problem doing.

In fact, they said they were willing to take a pay cut or move somewhere else if it meant that their job would offer them more opportunities for work-life balance, The Post reported.

Unfortunately, it's not something Millennial's are getting, *The Post* reported.

Wanting flexibility or work-life balance is the No. 1 thing we hear all the time from candidates. It's the No. 1 reason why people are looking for a new job, by far, Heidi Parsont, who runs TorchLight, a recruiting firm, told *The Post*. We're definitely seeing more candidates asking for it. But companies still see it as making an exception. It's still not the norm.

Corporate jobs that aren't fun

Millennial's don't want to get dressed in the corporate suit and tie anytime soon. A new survey from consulting firm Accenture found that just 15 percent of 2015 graduates want to work for a large corporation, CNN Money reported. Most youngsters prefer to work for medium-sized businesses that promote a fun work environment and atmosphere.

A full 60 percent of 2015 grads — and 69 percent of 2013 and 2014 grads, who were also surveyed — said they'd rather work for a company that has a 'positive social atmosphere' even if it means lower pay, CNN Money reported. Of course, they may think differently after a few years of working for the (low-paying) Man, and after Mom and Dad stop subsidizing them.

Any job with an employer

That's right. Millennial's are more likely to work by or for themselves. Inc. magazine reported in 2014 that 70 percent of Millennial's feel they'll work independently at some point in their careers. Similarly, Fast Company's Dr. Tomas Chamorro-Premuzic reported that Millennial's are the most likely generation to be self-employed.

This is for several reasons, including some that we covered above. Chamorro-Premuzic also believes that Millennial's underestimate the challenges of starting their own company, which makes them more willing to launch out on their own.

But the biggest reason many young adults want to work for themselves may have to do with the corporate and traditional jobs that many Millennial's have had in the past, according to Chamorro-Premuzic.

The No. 1 reason for this is that they have been traumatized by previous experiences with bosses, he wrote. We call them 'necessity entrepreneurs' but only because they have the necessity to avoid incompetent bosses. One cannot blame Millennial's for trying to do the same.[22]

---

[22] *6 jobs millennials don't want.* Retrieved from: http://national.deseretnews.com/article/5512/6-jobs-Millennial's-dont-want.html

In contrast to the Millennial profile, what is the Personality Profile of scientists and engineers?

A **scientist** is a person engaging in a systematic activity to acquire knowledge. In a more restricted sense, a scientist may refer to an individual who uses the scientific method. The person may be an expert in one or more areas of science. This article focuses on the more restricted use of the word. Scientists perform research toward a more comprehensive understanding of nature, including physical, mathematical and social realms....

Scientists are also distinct from engineers, those who design, build, and maintain devices for particular situations; however, no engineer attains that title without significant study of science and the scientific method. When science is done with a goal toward practical utility, it is called applied science. An applied scientist may not be designing something in particular, but rather is conducting research with the aim of developing new technologies and practical methods. – Wikipedia

And another pointed commentary on the relationship between temperament and profession:

The most important consideration in choosing one's life's work, and the one most frequently overlooked, is temperament. Each one of us is unique, but people do share some qualities of temperament, though to different degrees and in different combinations. These temperament qualities strongly influence our relationships, our preferences in life in many ways, the kind of work we enjoy, and the kind we do well.

On a fairly regular basis I give lectures on Entrepreneurship. I delight in telling audiences that I turned down a free Harvard Law School education. After seeing the look of amazement on the faces of the audience, I proceed to give a full account. My mother always wanted me to pursue this course based on the fact that - and this is probably a remark that only those of us who went through the Depression can understand – 'No Harvard Law School graduate ever had to walk the streets looking for a job.' Without complete conviction, after I got out of the Army in 1945, under very competitive circumstances, I applied and was admitted. Furthermore, based on the 'G.I. Bill', that is support for education for veterans, Uncle Sam would have paid for it. Then I throw in another shocker, 'And it was one of the best decisions I ever made.'

The point is that I would not have been a good lawyer. I am too restless and I get too many intervening thoughts to be good at research. Furthermore, lawyers spend much of their time involved in controversy, and I hate controversy. My abilities and enjoyment lie along the lines of imaginative new ideas, and imaginative and new ideas are not encouraged in the law. It is the old ones - the precedents - that count. Of course, it takes great skill to research preceding cases and to use the proper ones to prove one's case, but it is a different skill from dreaming up business ideas.

In a sense I lucked out because I was acting intuitively and without much guidance. Fortunately, there are ways, now, in which a person can analyze, or get help analyzing, one's temperament and take advantage of the direction which such an effort can yield in choosing a vocation.

A great help in this regard is the Myers-Briggs Type Indicator, which, through a series of simple questions of choice, identifies predominant characteristics and also the combination of characteristics for an individual. It categorizes people in terms of: 1) Extroversion vs. Introversion, 2) Thinking vs. Feeling, 3) Intuition vs. Sensation, and 4) Judging vs. Perceiving. Since there are four elements of choice, with two possibilities in each case, there are sixteen potential combinations. If you have never been introduced to this, it may sound complicated, but the test is very simple, as is determining each person's combination. The test results are followed by a description of each of the sixteen types, and when one finds the applicable type, genuine amazement usually results. Responses I have heard, as well as similar ones, are: 'Uncanny,' 'I can't believe they could learn so much about me,' 'He must have been following me around.'

One of the amusing results is that people usually like their types and think that everybody, if given a choice, would choose the same thing. I know of some engineers who felt that if others did not come out the same way, they 'flunked.'

You cannot flunk, as there is no passing grade. In learning about your temperament it is just a confirmation that you are a unique individual, a special person in your own right.

Temperament testing is helpful in choosing a vocation. For example, the test for Barbara, the social worker … who counsels pregnant teenage girls, indicated that people of her type '...make outstanding individual therapists.' Many similar situations could be cited. People have said, with surprise, as vocational activities are recommended based on their temperament, 'Of course, that is what I should be doing.'[23]

By contrast, the personality traits of engineers and scientists (according to test terminology: INTJ) are summarized below:

(Portions of the Scientist and Engineering profile that Millennial's may find irksome are highlighted.)

As an INTJ, your primary mode of living is focused internally, where you take things in primarily via your intuition. Your secondary mode is external, where you deal with things rationally and logically.

INTJs live in the world of ideas and strategic planning. They value intelligence, knowledge, and competence, and typically have high standards in these regards, which they continuously strive to fulfill. To a somewhat lesser extent, they have similar expectations of others.

---

[23] *The Influence of Temperament on Vocational Choice*, Chapter 29. Bryan Bell. Retrieved from: http://bbll.com/ch29.html

With Introverted Intuition dominating their personality, INTJs focus their energy on observing the world, and generating ideas and possibilities. Their mind constantly gathers information and makes associations about it. They are tremendously insightful and usually are very quick to understand new ideas. However, their primary interest is not *understanding* a concept, but rather *applying* that concept in a useful way. Unlike the INTP, they do not follow an idea as far as they possibly can, seeking only to understand it fully. INTJs are driven to come to conclusions about ideas. Their need for closure and organization usually requires that they take some action.

INTJ's tremendous value and need for systems and organization, combined with their natural insightfulness, makes them excellent scientists. An INTJ scientist gives a gift to society by putting their ideas into a useful form for others to follow. It is not easy for the INTJ to express their internal images, insights, and abstractions. The internal form of the INTJ's thoughts and concepts is highly individualized, and is not readily translatable into a form that others will understand. However, the INTJ is driven to translate their ideas into a plan or system that is usually readily explainable, rather than to do a direct translation of their thoughts. They usually don't see the value of a direct transaction, and will also have difficulty expressing their ideas, which are non-linear. However, their extreme respect of knowledge and intelligence will motivate them to explain themselves to another person who they feel is deserving of the effort.

INTJs are natural leaders, although they usually choose to remain in the background until they see a real need to take over the lead. When they are in leadership roles, they are quite effective, because they are able to objectively see the reality of a situation, and are adaptable enough to change things which aren't working well. They are the supreme strategists - always scanning available ideas and concepts and weighing them against their current strategy, to plan for every conceivable contingency.

INTJs spend a lot of time inside their own minds, and may have little interest in the other people's thoughts or feelings. Unless their Feeling side is developed, they may have problems giving other people the level of intimacy that is needed. Unless their Sensing side is developed, they may have a tendency to ignore details which are necessary for implementing their ideas.

The INTJ's interest in dealing with the world is to make decisions, express judgments, and put everything that they encounter into an understandable and rational system. Consequently, they are quick to express judgments. Often they have very evolved intuitions, and are convinced that they are right about things. Unless they complement their intuitive understanding with a well-developed ability to express their insights, they may find themselves frequently misunderstood. In these cases, INTJs tend to blame misunderstandings on the limitations of the other party, rather than on their own difficulty in expressing themselves. This tendency may cause the INTJ to dismiss others input too quickly, and to become generally arrogant and elitist.

INTJs are ambitious, self-confident, deliberate, long-range thinkers. Many INTJs end up in engineering or scientific pursuits, although some find enough challenge within the business world in areas which involve organizing and strategic planning. They dislike messiness and inefficiency, and anything that is muddled or unclear. They value clarity and efficiency, and will put enormous amounts of energy and time into consolidating their insights into structured patterns.

Other people may have a difficult time understanding an INTJ. They may see them as aloof and reserved. Indeed, the INTJ is not overly demonstrative of their affections, and is likely to not give as much praise or positive support as others may need or desire. That doesn't mean that he or she doesn't truly have affection or regard for others, they simply do not typically feel the need to express it. Others may falsely perceive the INTJ as being rigid and set in their ways. Nothing could be further from the truth, because the INTJ is committed to always finding the objective best strategy to implement their ideas. The INTJ is usually quite open to hearing an alternative way of doing something.

When under a great deal of stress, the INTJ may become obsessed with mindless repetitive, sensate activities, such as over-drinking. They may also tend to become absorbed with minutia and details that they would not normally consider important to their overall goal.

INTJs need to remember to express themselves sufficiently, so as to avoid difficulties with people misunderstandings. In the absence of properly developing their communication abilities, they may become abrupt and short with people, and isolationists.

INTJs have a tremendous amount of ability to accomplish great things. They have insight into the Big Picture, and are driven to synthesize their concepts into solid plans of action. Their reasoning skills gives them the means to accomplish that. INTJs are most always highly competent people, and will not have a problem meeting their career or education goals. They have the capability to make great strides in these arenas. On a personal level, the INTJ who practices tolerances and puts effort into effectively communicating their insights to others has everything in his or her power to lead a rich and rewarding life.[24]

Downsides for Millennial's other than those highlighted above, mostly drawn from my experiences as a professional.

1. Science and engineering have mostly moved far beyond the garage workshop. Competitive innovation and invention is expensive. So these innovators work for a University or large

---

[24] *Portrait of an INTJ - Introverted iNtuitive Thinking Judging*
(Introverted Intuition with Extraverted Thinking). The Scientist. Retrieved from:
https://www.personalitypage.com/INTJ.html

Private or Governmental research firm.

2. Millennial's often see in the media individual scientist's pictures and credits. And so see Science as an isolated, individual endeavor. Not always so.

3. At all levels quoted above, there may have been an "originator" of the idea one is working on (such as artificial nuclear fusion), but to bring anything so complicated into reality takes many participants and often many, many years.

Example: Forty years ago, as a graduate student, I worked in a group with two Physicists, several Post-doctoral fellows, and several fellow graduate students. The research I did for my dissertation was proved to be inadequate to do the job a few years after I graduated and the area went dead for two decades. I entered that particular research because I could foresee the energy crunch we are now in, and above all, thought free energy might help prevent war.

However, with the building of the first *production* fusion reactor, ITER[25], my original research was a key factor in the design of the injector gun. That was serious delayed satisfaction! And there never was a promise of long term self-satisfaction at all; one may be working in a blind alley.

4. Innovators at all levels *often* have to stand alone, since there are long traditions and change is often resisted. Innovators must praise themselves, not having someone watching over their shoulders and praising every move. The praise comes from the team when the work is *finished*.

5. The current generation of Scientists are frequently seriously introverted, and have poor social skills. They can be difficult to deal with on the personal level.

6. Scientists frequently work far beyond 40 hours per week, as

---

[25] ITER ("The Way" in Latin) is one of the most ambitious energy projects in the world today. Retrieved from: http://www.iter.org/

noted above.

7. Today's scientists in-training have been let down seriously by their K-12 schools. They enter College or University with barely a sixth-grade education. Remediation is required (which attacks the ego), and Science degrees take longer than normal to complete.

America's Failed Public Education System Undermining American's seeking a Degree

There was a time several decades ago when Higher Education in America was the best in the World. Well to do parents in other countries sent their children here to College/University.

Of course, there was a preliminary step for Americans: the K-12 education, which functioned effectively in the past so as to prepare Americans for College/University.

Neither of the statements is true any longer.

The World Education ranking is shown below:[26]

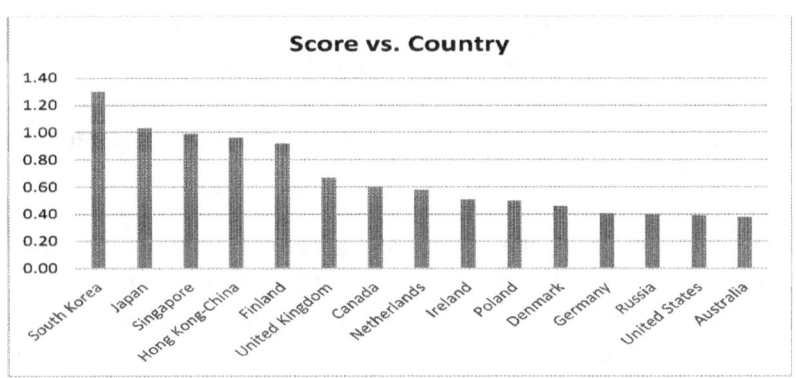

[26] Index - Which countries have the best schools? Pearson. Retrieved from: http://thelearningcurve.pearson.com/index/index-ranking. Our Index, which uses global data sets such as PISA, TIMSS and PIRLS together with individual country data such as literacy and graduation rates.

What has happened over time?

Parental hostile take-over of K-12 school system.

Issues:

A. Control of topics taught in public school. Parents were in the past as well as the present concerned about schools teaching sex education and evolution theory.
B. Their children's self-esteem. Following the Great Depression, there was societal depression. Living had been HARD! Parents didn't want to have THEIR children experience such hardship, and so abundantly showered their children with (often unearned) "things" and "praise". Psychology was also going through one of its fads: the concern that judging people leads to poor self-image leads to acting out. For instance, not scoring sporting events.

    When students first enter the real world work-place, they feel entitled to praise and rewards, regardless of their performance. And maybe this results in the first punishment they have ever had, firing.

C. It is common for Millennials to change jobs frequently in the first few years, seeking the ideal job that doesn't exist.

Politicizing education became a football to divert from other egregious government actions

There has been a decline in public school performance, although possibly with cause. As an educator myself, the skills required in the work-a-day environment have been increasing geometrically as we have progressed technologically. Much more real-life knowledge is required of today's worker than when I entered my first job. There is probably justifiable resistance to ever-on-the-move and increasing standards.

Political Posturing:

- Politicians Decided to Make Teachers "Accountable" for their Student's Outcomes.
- They blamed the Teacher's Unions for protecting "bad" teachers (only somewhat true).
- They believed that assessing student learning is easy: test a LOT! Actually, assessing learning is VERY difficult.
- Teachers began spending all their time "teaching to the test", with very little time left to teach actual workforce skills.
- One of my former students said "No Child Left Behind", left everyone behind.
- Someone decided that every child should go to college, rather they had aptitude or interest in doing so

American social attitudes have changed

I was shocked a decade ago to attend a National Science Foundation Grant Writing Workshop, and heard a very not-politically correct statement: American individuals of third generation (and on) have taken for granted the hard work that made America what it is. Youth have adopted the maximum-outcome/minimal-effort mindset, and stopped going into the occupations that require the most student buy-in: the Sciences.

This statement was made before the Millennial's came along, but has a similar theme: times have been good, and Millennials feel entitled to the good life regardless of their efforts at work.

As shown above, it may not be laziness but personality mismatch.

Results:

- It is a fact that one-quarter of America's Physicians today are foreign-born[27]

---

[27] *Foreign-Born Health Care Workers in the United States*. Migration Policy Institute. Retrieved from: http://www.migrationpolicy.org/article/foreign-born-health-care-workers-united-states/

- The science, technology, engineering and mathematics (STEM) fields are seriously understaffed, and America does not have a robust pipeline of trainees in STEM areas (see Chapter 1).

Solutions:

To attack with these shortfalls in the STEM pipeline the US Government has moved in several directions:

Solution 1: The Federal Government is targeting immigrants from Latino Countries where "economic comfort" and "gadgets" have not yet set in.

> It is estimated that <u>Hispanics will comprise 30 percent of the U.S. population by 2040</u> and will be the majority group in several states (U.S. Census Bureau, 2008).[28]
>
> Science, Technology, Engineering, and Mathematics skills are necessary now more than ever in order to compete in a global economy. According to the U.S. Congress Joint Economic Committee (JEC), between 2010 and 2020 the <u>overall employment in STEM occupations will increase by 17 percent</u>, yet not enough students are pursuing degrees and careers in the STEM fields to meet the increasing demand. There are currently two science and technology job openings for every qualified job seeker.
>
> The lack of STEM representation is even more prevalent among Hispanics, who although accounted for 16% of the U.S. population in 2010, only earned 8 percent of all certificates and degrees awarded in the STEM fields between 2009 and 2010.[29]

---

[28] *Overview of Hispanics in Science, Mathematics, Engineering and Technology (Stem): K-16 Representation, Preparation and Participation.* Gloria Crisp. Retrieved from: http://www.hacu.net/images/hacu/OPAI/H3ERC/2012_papers/Crisp%20nora%20-%20hispanics%20in%20stem%20-%20updated%202012.pdf

[29]*Hispanic and STEM Education.* US Department of Education. Retrieved from: http://www2.ed.gov/about/inits/list/hispanic-initiative/stem-factsheet.pdf

In 2013, the <u>University STEM Faculty population</u> by ethnicity was[30]:

Hispanic	White	African American	Asian
5%	78%	6%	10%

The 2010 <u>University STEM Student population</u> by ethnicity was[31]:

Discipline	Hispanic	White	African American	Asian
Computer Science	7.4%	67%	11.5%	8.5%
Engineering	6.9%	69.5%	4.4%	12.4%
Mathematics	6.4%	71.8%	5.3%	10.4%
Physical Sciences	5.5%	73.7%	5.5%	10.5%

Solution 2: There is also a pipeline of International Students 2013 (some of these are hidden in the above Table).[32]

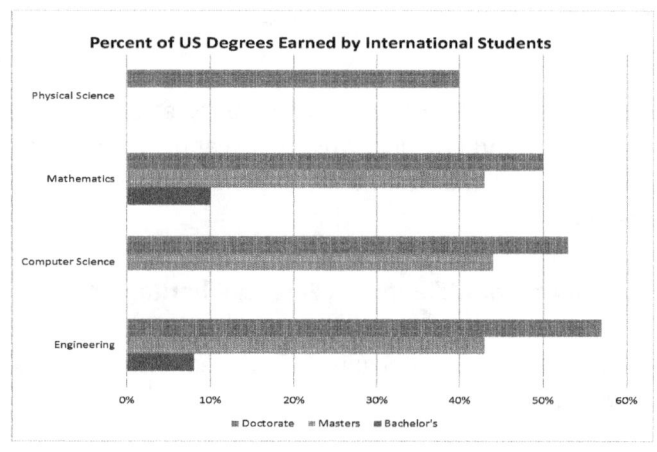

[30]*Race/ethnicity of college faculty*. National Center for Educational Statistics. Retrieved from:
https://nces.ed.gov/fastfacts/display.asp?id=61

[31] http://www.directemployers.org/2012/08/16/the-college-class-of-2013-current-demographics/
[32] Pew Research Organization.

The hope is that America can retain many of these students after graduation.

Solution 3: Importing degreed professionals was a robust program before 911. The history is shown below.[33]

I don't know *which* professionals were being given these EB-1 visas, but the general trend is there.

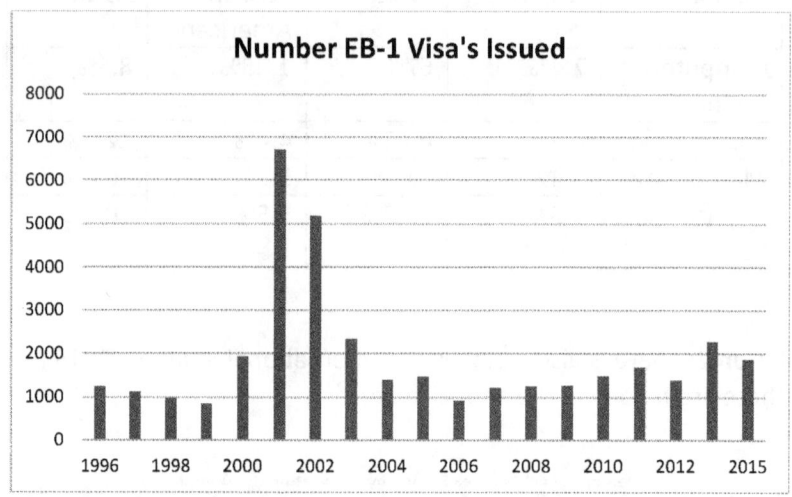

This plan was restricted by Homeland Security after 911 when it was becoming effective and growing. Now this avenue seems to be opening again. *This is probably the best method of making supply meet demand for professionals.*

A final impediment for Millennial's: Financial heritage

In the 1980's, President Reagan introduced Reaganomics to pull America out of an economic slump:

Buy now, pay later.

[33] *Visa Statistics*. US Department of State. Retrieved from:
https://travel.state.gov/content/visas/en/law-and-policy/statistics.html

This plan began American Individuals (and the American Governments) to begin living in the future. The only cost was loan interest!

However, that interest has America trapped now. Instead of pay-as-you-go and balancing via income and outflow "necessity", individuals and governments racked up huge debt.

For families, it became necessary for both parents to work to pay their own interest and to pay towards the government's interest via increased personal income tax.

The American family changed character, and exposed children to unsupervised, alone time.

Latchkey kids are kids between the ages of 5 and 13 who take care of themselves with no adult supervision before and after school on a regular basis. The term self-care is also used to describe these children.

Although children from single-parent working families and lower income children do spend time home alone, children from higher incomes actually spend more hours alone. Research indicates one reason for this could be that higher income neighborhoods might feel safer to parents.

Unsupervised time can make children independent and sometimes can lead to challenges and negative outcomes. School performance tends to be lower in latchkey children. Some studies show the number of latchkey students doing poorly in school is as high as 51%. Children tend to go home and watch television rather than engage in more stimulating activities. Research by the National Educational Longitudinal Study found that the amount of television time could be predicted by the number of self-care hours. After-school programs have significantly improved school performance because they often provide homework help.

Drug use is higher among latchkey kids. Students unsupervised after school have a higher rate of trying drugs than those that are supervised. Research by the University of California and other universities found a relationship between the amount of self-care and drugs abuse in 8th graders regardless of other factors such as income and extracurricular activities. Research by Laurence Steinburg found that peer pressure, which can lead to drug use, was higher when students were in a public place, like malls, rather than in homes, so students who hang out with their friends instead of going home are at risk.

Sexual activity is also higher when adolescents are not supervised after school....

Latchkey children tend to be more obese because they tend to stay in the house when home alone instead of playing outside. In addition, they may be more likely to have an easy unhealthy snack instead of preparing a healthy snack after school.[34]

Somewhere along the way, the "happiest place on Earth" (America) became unhappy, and people felt trapped. Drug usage has become the common salve for unhappiness for vast numbers of Americans.

Another source of increased crime!

---

[34] *Latchkey Kids: Definition, Effects & Statistics*. Retrieved from:
http://study.com/academy/lesson/latchkey-kids-definition-effects-statistics.html

CHAPTER 3 EXCEPTIONS TO THE MILLENNIAL PROFILE

There is a significant segment of the Millennial Generation that does not fit the Millennial profile.

Illustration: I was sitting on an interdepartmental program group before the session began. The group was to address, among other things, how to motivate students to finish their college programs and graduate.

Over and over, my fellow faculty said, "I went through everything agonizing that is involved in earning University degrees, not so much because of what I wanted, but because of what I DID NOT WANT".

Mostly what they did not want was the Hell of poverty in which they were raised, the vulnerability of their family to external events, the family addictions and insanities, etc. Or being stuck in a going-nowhere, boring job.

In my case, my early family life was in a working-poor family. When I was about 10 years old, my father had a heart attack, the medical bills stacked up (minimal health insurance through his union), there was a down-time to recover with no income, and he had to change to a less stressful profession. It took multiple moves to accomplish that. The whole process put my family flat on its butt.

My father always came home so exhausted that he ate a light dinner and either parked on the sofa or just went to bed. There was little time for me, but I understood: he was keeping me from being hungry and homeless.

Then, when I was about 14 years old, my Mother became ill with Cancer. Same basic pattern as above. My college fund (from working as a teen) went to her medical bills. Family flat on its butt.

I was bright, but no funds to attend college. Fortunately, a scholarship came out of the blue and I finished through my BA.

As a budding scientist during the Cold War, the Federal Government supported me through graduate school.

I thought I was home free.

(Nevertheless, at least twice serendipity hit me, and I was twice flat on my butt.)

Students who have been raised in homes with relative comfort and few, if any, bumps in the road are the Millennial's described in the profile above.

But for the rest of us, we will bust our humps to get away from poverty and vulnerability and exposure.

So in undergraduate school (almost 50 years ago), when my classmates partied the whole weekend, I was studying. Subjects did not come easily to me, but I found time-on-task made me as good as or better than my classmates.

Students with such early family experiences can still jump the gap to Upper Middle Class.

Even after graduation, the fight isn't over. These students will still have blue collar characteristics (such as speaking their minds, rather than flow with the crazy social mores in the white collar arena). It will take at least three consistent generations of effort to "fit" into their new class with no "tells".

If you do have a "tell" about your past, the White Collar people around you will know and attempt to take advantage of you. They know you have worked extraordinarily hard in the past, and will give you the most difficult and irksome assignments.

When that White Collar arrogance and privilege and dominance has been used against me, my health deteriorates. I quit and move on.

My cousin says, "Well, I've been looking at these assholes long enough. I will change jobs and look at a new set of assholes".

But, if recognized, our history gives us greater strength than our peers, and we can plough ahead with confidence to a solution.

I see this every semester I teach college/university courses. Those students who have lived through a rough childhood have learned to be organized and take care of whatever comes their way. They have grown up early and have adult skills far beyond their classmates. These are the ones who persist and attain their goals.

CHAPTER 4 AMERICA THE VULNERABLE

What kind of American are we leaving to the Millennial's?

Over the years, the wealthy must be exchanging LMAO notes. They support the Federal Government to babysit the "genius nursery".

Yet, the Federal Government is completely incompetent to manage or to motivate anyone.

- Congress is egocentric. They have consistently approved programs that are not paid for. They either borrow the money (the National Debt) in a way that no sane business person would do, or else they build Ponzi schemes. Present workers are paying for benefits for people who paid into the system long ago; those who are taking out of the system paid for by workers of the last generation.

If each program had been paid for at inception, then these programs would have been pay-as-you-go, as any private investment program operates.

- I know of one case where Congress legislated measurement units so that a Law of Nature (Ohm's Law) was no longer true. Congress's response was that they were in charge of law making. Somehow they couldn't distinguish between immutable laws of nature, and their power to legislate behavioral *incentives* on the fly.

- There is a critical issue constantly in debate between the Federal Government parties on the basis of what they can do: is the constitution (written 300 years ago, within a cultural and technological setting) ABSOLUTE, or is it a LIVING set of guidelines that needs occasional updating.

This same debate occurs with religions. The issue is confidence versus fear.

A definition of intelligence is:

The ability to tolerate ambiguity. Intelligent people are confident enough to see the world as grey, not black-and-white, and willing to take on the process of slugging through a "new" situation.

Fearful and unsure people don't want to think critically (an emotional and physical strain and drain), and relax in laziness back to a black-and-white document of the past.

National Debt

As mentioned above, the Federal Government has allowed an approach to an unacceptable debt, which a business manager would never do. The risk of collapse is too great.

> The public debt is defined as how much a country owes to lenders outside of itself. These can include individuals, businesses and even other governments. The term 'public debt' is often used interchangeably with the term sovereign debt....

> Public debt is the accumulation of annual budget deficits. It's the result of years of government leaders spending more than they take in via tax revenues.

> On January 29, 2016, the U.S. debt surpassed $19 trillion. That meant the debt-to-GDP ratio is 106%[35].... However, the public debt was a more moderate $13.7 trillion. That made the public debt-to-GDP ratio a *safe* 76%. According to the World Bank, the tipping point is 77%.[36]

---

[35] The debt-to-GDP ratio compares a country's sovereign debt to its total economic output for the year. Its output is measured by GDP, or Gross Domestic Product. This ratio is a useful tool for investors, leaders, and economists. It allows them to gauge a country's ability to pay off its debt. A high ratio means a country isn't healthy enough to pay off its debt. A low ratio means there is plenty of economic output to make the payments.

[36] *What Is the Public Debt? Pros, Cons, and How to Tell When It's Too High.* Retrieved from: http://useconomy.about.com/od/usdebtanddeficit/p/Public-Debt.htm

Personal Debt

For many American households, the recession was a time to pay off debt and get their finances in order—whether they wanted to or not. But according to the latest data from the Federal Reserve's Flow of Funds (PDF), Americans are taking on debt once again. The difference is that this time we're borrowing to finance new cars, college tuition, and other consumer goods.

American household debt peaked in 2007 and has since fallen 15 percent. Home mortgage debt accounted for much of the decline—it's dropped 22 percent since 2007. Consumer debt, on the other hand, has continued to increase and just reached an all-time high of $3.2 trillion.

Americans have added about $100 billion of student debt a year to their balance sheets since 2008. Credit cards and auto loans have also come roaring back, particularly auto loans. The amount of outstanding auto debt is the highest it's ever been.

Auto and credit card debt, while overall much smaller than student or mortgage loans, is in some ways more risky. Student loans and mortgage debt both finance an asset that's expected to increase in value. A mortgage finances real estate, which has good odds of increasing in value or, at least, holding your housing costs stable for 30 years. Student debt is an investment in future earnings. There's no guarantee, but the odds are if you finish college, your salary will, over time, recoup your investment. So while the explosion in student debt and the rising delinquency rate are troubling, at least some of the debt can be justified: It's a leveraged investment that has a decent chance of paying off.

That's often not the case for auto loans and goods bought with a credit card. A reliable car may be a necessary expense for some, but as an asset it's guaranteed to depreciate. Beyond safety and reliability, there's little investment value in buying a new car.

It seems the increase in auto spending wasn't driven solely by Americans buying the cheapest, safest car they could find.

According to the Bureau of Economic Analysis, spending on cars has increased 35 percent since the recession, almost all on new cars. Spending on repairs and net used cars has barely budged. The surge in new-car buying is partially because households that cut back on big-ticket items during the recession are spending again. But the fact that spending seems to be coming at the expense of more debt suggests Americans are putting themselves in a riskier financial position. They may have less debt overall, but an increasing share of that debt finances consumption that only declines in value.

Consumption per capita has been rising since the recession, despite stagnant income. This may revive demand for now, but the financial crisis showed that consumption, financed by debt, is not the path to resilient growth.[37]

Social Security and Medicare

People have for decades paid into these Federal Government programs, and now expect the returns guaranteed by the Federal Government. Of course, this is a Ponzi scheme, but who created it? The Federal Government.

When push comes to shove, will the Federal Government default on its foreign creditors, or on its own citizens? Either is a failure of government.

Who owns the National Debt?

As of September 2014, foreigners owned $6.06 trillion of U.S. debt, or approximately 47% of the debt held by the public of $12.8 trillion and 34% of the total debt of $17.8 trillion. The largest holders were China, Japan, Belgium, the Caribbean banking centers, and oil exporters. The share held by foreign governments has grown over time, rising from 13% of the public

---

[37] *Consumer Debt Hits an All-Time High*. Blomberg Business. Retrieved from:
http://www.bloomberg.com/bw/articles/2014-09-30/consumer-debt-hits-an-all-time-high

debt in 1988 to 25% in 2007.[38]

At least with China, we have now lost bargaining power in international financial matters.

Where are America's Manufacturing Facilities?

Business over-taxation, in general and specifically manufacturing, procedures has led to businesses both outsourcing and offshoring.

They were forced into a corner.

I remember when the State of California made the same mistake, and their technology businesses fled the State, mostly to Denver and Atlanta.

Fortunately, these jobs are gently coming back to America.[39]

But now there is renewed interest in increasing individual and corporate taxes to make up for Congress's mismanagement. This will bring about another egress of corporations.

From my viewpoint, suppose we have to go to war with China. Will we ask them to equip our troops with gear and weapons if we have given over control of *that* capability to China? And what is to prevent China, say in an economic slump of their own, from Nationalizing American holdings in China?

US Trade Deficit

Our global economy runs on the expansion of credit. Since 1970, the credit markets have doubled in size every decade. This massive expansion of debt/borrowing created one of the biggest booms in history - seen in the DOW running from under 1000 in

---

[38] *Consumer Debt Hits an All-Time High. Bloomberg Business*. Retrieved from: http://www.bloomberg.com/bw/articles/2014-09-30/consumer-debt-hits-an-all-time-high

[39] *Made in USA (Again): Why Manufacturing Is Coming Home*. Retrieved from: http://www.inc.com/eric-markowitz/the-long-journey-home-why-manufacturing-is-returning-to-the-usa.html

1980 to over 14,000 in the year 2000, and the real estate markets soaring in prices worldwide from 1997 to 2008 ....

The economy is now dependent on this growth of credit in order to expand in size. Why? Because you need a larger economy tomorrow to pay off the debt you accumulated yesterday. Does this sound like a Ponzi scheme? Well, it is.

This is why central banks and governments around the world are desperate to pump credit into the system through government spending, which has been paid for with central banks printing money.

Up until 2008, this credit expansion was handled through the consumers around the world ready and willing to take on more credit card debt, a new auto loan, or a home equity line. Then, we ran into a problem. When the subprime mortgage crisis hit in 2008 and home prices began to contract, the American consumer and engine of global growth hit the wall. Government leaders went to work to get the credit machine growing again and it appears, based on the most recent data, that they have accomplished this goal.[40]

Shown below is the balance of American trade in Exports and Imports historically. The vertical distance between lines is what America OWES to other countries in $Billions:[41]

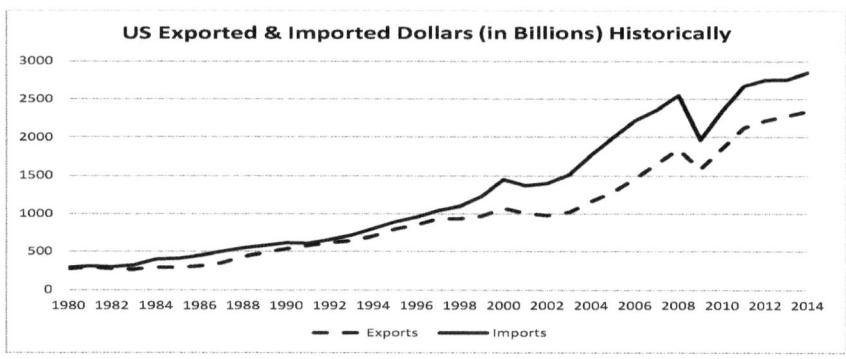

[40] U.S. Census Bureau, Economic Indicator Division. Information on data sources and methodology are available at http://www.census.gov/foreign-trade/guide/sec2.html#bop.
[41] Ibid.

Here is another source of foreign debt.

In 2014, we owed $509,000,000.

CHAPTER 5 THE CURSE OF MIDDLE AMERICA: *CONSUMERISM*

Affluenza: A Disease

Affluenza, a portmanteau of *affluence* and *influenza*, is a term used by critics of consumerism. It is thought to have been first used in 1954 but it gained legs as a concept with a 1997 PBS documentary of the same name and the subsequent book, *Affluenza: The All-Consuming Epidemic* (2001, revised in 2005, 2014). These works define affluenza as 'a painful, contagious, socially transmitted condition of overload, debt, anxiety, and waste resulting from the dogged pursuit of more.' The term 'affluenza' has also been used to refer to an inability to understand the consequences of one's actions because of financial privilege, notably in the case of Ethan Couch. – Wikipedia

And:

Affluenza's costs and consequences are immense, though often concealed. Untreated, the disease can cause permanent discontent. In our view, the affluenza epidemic is rooted in the obsessive, almost religious quest for expansion that has become the core principle of what is called the American dream. It's rooted in the fact that our supreme measure of national progress is that quarterly ring of the cash register we call the gross domestic product. It is rooted in the idea that every generation will be materially wealthier than its predecessor and that somehow, each of us can pursue that single-minded end without damaging the other countless things that we hold dear.[42]

Biology Lessons: Finite Space and Resources

In July 1968 four pairs of mice were introduced into the Utopian universe. The universe was a 9-foot (2.7 m) square metal pen with 4.5-foot-high (1.4 m) sides. Each side had four groups of four vertical, wire mesh 'tunnels.' The 'tunnels' gave access to nesting boxes, food hoppers, and water dispensers. There was no shortage of food or water or nesting material. There were no predators. The only adversity was the limit on space.

Initially the population grew rapidly, doubling every 55 days. The population reached 620 by day 315, after which the population growth dropped markedly. The last surviving birth was on day 600. This period between day 315 and day 600 saw a breakdown in social structure and in normal social behavior. Among the aberrations in behavior were the following: expulsion of young before weaning was complete, wounding of young, increase in homosexual behavior, inability of dominant males to maintain the defense of their territory and females, aggressive behavior of females, passivity of non-dominant males with increased attacks on each other which were not defended against. After day 600, the social breakdown continued and the population declined toward extinction. During this period females ceased to reproduce. Their male counterparts withdrew completely, never engaging in courtship or fighting. They ate, drank, slept, and groomed themselves – all solitary pursuits. Sleek, healthy coats and an absence of scars characterized these males. They were dubbed 'the beautiful ones.' Breeding never resumed and behavior patterns were permanently changed. – Wikipedia

---

[42] *Affluenza*, DeGraaf, Wann, & Thompson. Barret-Koehler Publishers, Inc. (2014).

Applied to Humans

All of our basic resources, such as land, water, energy, and biota, are inherently limited. As human populations continue to expand and finite resources are divided among increasing numbers of people, it will become more and more difficult to maintain prosperity and a quality of life, and personal freedoms will decline.

During recent decades there has been a dramatic worldwide population increase. The U.S. population doubled during the past 60 years from 135 million to more than 270 million and, based on the current U.S. growth rate of approximately 1% per year, is projected to double again to 540 million in the next 70 years. China's population is 1.3 billion and, despite the governmental policy of permitting only one child per couple, it is still growing at an annual rate of 1.2%.

India has nearly 1 billion people living on approximately one-third of the land of either the United States or China. India's current population growth rate is 1.9%, which translates to a doubling time of 37 years. Together, China and India constitute more than one-third of the total world population. Given the steady decline in per capita resources, it is unlikely that India, China, and the world population in total will double.

Thus far, the relative affluence enjoyed by most Americans has been possible because of an abundant supply of fertile cropland, water, and fossil energy. As the U.S. population continues to expand, however, resource shortages similar to those now being experienced by China and other developing nations will become more common. Accelerated declines in the U.S. standard of living are likely if the U.S. population increases as projected during the next 70 years, from 270 million in 1998 to 540 million.[43]

---

[43] *Will Limits Of The Earth's Resources Control Human Numbers?* Retrieved from: http://dieoff.org/page174.htm

Consumerism as a Way of Life

Definitions:

> 1. The practice of conspicuously buying flashy, expensive items and expecting them to fill the voids in one's life.

> 2. A pejorative used to malign people who enjoy material luxuries, implying that it's impossible for a person do so without having an unhealthy fixation on them. Usually used as a sour grapes excuse to hate people who can afford more luxuries than one's self.[44]

Excessive Consumerism

Excessive consumption leads to bigger houses, faster cars, trendier clothes, fancier technology, and overfilled drawers. It promises happiness, but never delivers. Instead, it results in a desire for more... a desire which is promoted by the world around us. And it slowly begins robbing us of life. It redirects our God-given passions to things that can never fulfill. It consumes our limited resources.

Consider this list of ten practical benefits of escaping excessive consumerism in your life:

Less debt. The average American owns 3.5 credit cards and $15,799 in credit card debt... totaling consumer debt of $2.43 trillion in the USA alone. This debt causes stress in our lives and forces us to work jobs that we don't enjoy. We have sought life in department stores and gambled our future on the empty promises of their advertisements. We have lost.

Less life caring for possessions. The never-ending need to care for the things we own is draining our time and energy.

---

[44] Urban Dictionary. Retrieved from:
http://www.urbandictionary.com/define.php?term=consumerism

Whether we are maintaining property, fixing vehicles, replacing goods, or cleaning things made of plastic, metal, or glass, our life is being emotionally and physically drained by the care of things that we don't need—and in most cases, don't enjoy either. We are far better off owning less.

Less desire to upscale lifestyle norms. The television and the Internet has brought lifestyle envy into our lives at a level never before experienced in human history. Prior to the advent of the digital age, we were left envying the Jones' family living next to us—but at least we had a few things in common (such as living in the same neighborhood). But today's media age has caused us to envy (and expect) lifestyle norms well beyond our incomes by promoting the lifestyles of the rich and famous as superior and enviable. Only an intentional rejection of excessive consumerism can quietly silence the desire to constantly upscale lifestyle norms.

Less environmental impact. Our earth produces enough resources to meet all of our needs, but it does not produce enough resources to meet all of our wants. And whether you consider yourself an environmentalist or not, it is tough to argue with the fact that consuming more resources than the earth can replenish is not a healthy trend—especially when it is completely unnecessary.

Less need to keep up with evolving trends. Henry David Thoreau once said, "Every generation laughs at the old fashions, but religiously follows the new." Recently, I have been struck by the wisdom and practical applicability of that thought whether relating to fashion, decoration, or design. A culture built on consumption must produce an ever-changing target to keep its participants spending money. And our culture has nearly perfected that practice. As a result, nearly every year, a new line of fashion is released as the newest trend. And the only way to keep up is to purchase the latest fashions and trends when they are released... or remove yourself from the pursuit altogether.

Less pressure to impress with material possessions. Social scientist Thorstein Veblen coined the phrase *"conspicuous consumption"* to describe the lavish spending on goods and services acquired mainly for the purpose of displaying income or wealth. In his 1899 book, *The Theory of the Leisure Class,* this term was used to describe the behavior of a limited social class. And although the behavior has been around since the beginning of time, today's credit has allowed it to permeate nearly every social class in today's society. As a result, no human being (in consumption cultures) is exempt from its temptation.

More generosity. Rejecting excessive consumerism always frees up energy, time, and finances. Those resources can then be brought back into alignment with our deepest heart values. When we begin rejecting the temptation to spend all of our limited resources on ourselves, our hearts are opened to the joy and fulfillment found in giving our personal resources to others. Generosity finds space in our life (and in our checkbooks) to emerge.

More contentment. Many people believe if they find (or achieve) contentment in their lives, their desire for excessive consumption will wane. But we have found the opposite to be true. We have found that the intentional rejection of excessive consumption opens the door for contentment to take root in our lives. We began pursuing minimalism as a means to realign our life around our greatest passions, not as a means to find contentment. But somehow, minimalism resulted in a far-greater contentment with life than we ever enjoyed prior.

Greater ability to see through empty claims. Fulfillment is not on sale at your local department store—neither is happiness. It never has been. And never will be. We all know this to be true. We all know that more things won't make us happier. It's just that we've bought into the subtle message of millions upon millions of advertisements that have told us otherwise. Intentionally stepping back for an extended period of time helps us get a broader view of their empty claims.

Greater realization that this world is not just material. True life is found in the invisible things of life: love, hope, and faith. Again, we all know there are things in this world that are far more important than what we own. But if one were to research our actions, intentions, and receipts, would they reach the same conclusion? Or have we been too busy seeking happiness in all the wrong places?[45]

I will add: **The Merry-Go-Round**

Key Word: Planned obsolescence[46]:

1. A product is produced that is necessary for life of the common individual, or has been *promoted* to be necessary. It is constructed poorly or in such way that the consumer becomes bored with its simplicity.
2. It wears out, or a flashier product appears on the market. Also destined for obsolescence.
3. The manufacturer has a steady income.
4. The product is produced offshore or sublet overseas.

---

[45] *10 Reasons to Escape Excessive Consumerism*. Retrieved from:
http://www.becomingminimalist.com/escaping-excessive-consumerism/

[46] **Planned obsolescence** or **built-in obsolescence** in industrial design and economics is a policy of planning or designing a product with an artificially limited useful life, so it will become obsolete, that is, unfashionable or no longer functional after a certain period of time. The rationale behind the strategy is to generate long-term sales volume by reducing the time between repeat purchases (referred to as "shortening the replacement cycle").

5. The major cost in dollars of the product leaves America.
6. America owes a foreign debtor.
7. The manufacturer advertises something flashier
8. The addicted consumer buys.

Example: one telephone company charges a flat rate for service. Included is a "new iPhone" every time one is released. Do you really need "the latest" and do you know what you are paying for?

"Round and round she goes, and where she stops nobody knows".[47]

Does Anyone Win?

In their powerful book, *The Spirit Level*, the British epidemiologists Richard Wilkerson and Kate Pickett show highly unequal countries have poorer outcomes in more than a dozen indicators of wellbeing, from health to happiness to crime.

While in every country, the rich are happier and healthier than the poor, even the rich in unequal countries are not as healthy or happy as those in more egalitarian ones. Wealthy Americans live only about as long as poor Europeans.[48]

And

Nevertheless, America's obsession with wealth continues unabated as we continue to pursue the chimera that freedom means no limits in the right to get as rich as possible.[49]

---

[47] From The Original Amateur Hour and later game shows.
[48] Poverty. The World Bank Group last updated April 2013. http://go.worldbank.org/VL7N3V6F20.
[49] *Affluenza*, DeGraaf, Wann, & Thompson. Barret-Koehler Publishers, Inc. (2014).

CHAPTER 6 WRAPPING UP

So let's sum up:

The American Economy, as it currently is, requires a population of innovators and inventers to satisfy the consumer (who is comfortable financially, and bored) with "new" products and "ideas" every couple years.

The Top 2% of the wealthy pay for one-half the costs to operate the Federal Government; they continue to do so in order maintain a level of safety (at home and from abroad), and as caretakers of the "idea nursery". The necessary and cherished population.

The Federal Government is so tied up in themselves and their re-election or job permanence, that the Government at the top levels is grid-locked. They are no longer "watching" the cherished population, and currently that population is diminishing or constant. Constant means we all lose our ability to grow our economy.

Support to the National Science Foundation is going down (apparently seen as a luxury) and support to the National Institutes of Health have increased (apparently the promise of better health and extended life-times Congress feels is a necessity).

The past generation of adults in the Middle Class have worked hard, built a comfortable foundation to life, and (largely) produced the current generation entering the work place: the Millennial's.

Most of the Millennial's haven't seen much lack of goods and services, take for granted what their parents and grandparents busted their humps to build, and have made a number of personality and life-style choices that will have consequences.

Personally, I would love to have had the millennial outlook during my nearly finished life-time, but I never had the luxury to do so.

Thus, the life-style choices Millennial's have made draw them into jobs that are: service, low responsibility, and low participation. And these jobs occur *now* in the Lower Middle class.

That leaves a vacuum in the Upper Middle Class.

There are three sources (that I know of) to replace that population.

I don't know if this is intelligent design, the free-market, or fortunate serendipity.

Those sources are:

- America's poor and working poor (the American Dream is still there)

- New immigrants from third world countries with a strong (not diluted) work-ethic

- Purposefully immigrating degreed people from abroad.

These sources will gradually replace the Upper Middle Class.

Note: actually, as I researched this book, I found many disciplines outside the STEM area that have the same needs for "trained" personnel in order to maintain, or to grow.

Already, in my personal interactions, I see the influx of this new class into the Health Care and Education professions.

This is the reorganization of the Middle Class, and some of the players who brought it about.

It is amusing to me what the history has been. There is a concept from Quantum Mechanics: The Uncertainty Principle. This Principle is only valid at the elemental particle level, and it states that one cannot simultaneous know exactly the position and the speed of an <u>elemental particle</u>.

I illustrate this Principle in class at the macroscopic level by talking about determining the exact position of a wet watermelon seed. The more one tries to localize it in one dimension exactly, by squeezing down hard, the more likely it is to squirt horizontally in the transverse direction.

We are following a generation when parents and politicians tried to control the Millennial's into the Upper Middle Class, but those exact actions destined the Millennials into the lower Middle Class.

It is easy to see this in hindsight, and I don't claim a conspiracy.

Good people did what they thought was right: unfortunately, they forgot who the economic Golden Goose is, and killed it.

It is a tragedy that we have produced a generation prepared to live a comfortable, trouble-free, Eden-like life, when America has nothing but Hellish problems ahead:

- ➢ Individuals and American Governments deeply in debt

- ➢ Natural resources running out

- ➢ Rampant consumerism

- ➢ Population (consumers) growing geometrically

- ➢ America virtually without manufacturing capability

- ➢ Paralyzed Federal Government

- ➢ Failed Education System.